The Smart Product Manager's Guide to Connectivity in the Packaging Industry

Learn what's important what to do next in the ever-changing world of the digital factory. End the uncertainty of your product road map

JOHN RINALDI

Copyright © 2019 Real Time Automation, Inc.

All rights reserved.

ISBN: 9781686308130

PACKAGING PRODUCT MANAGER'S CONNECTIVITY GUIDE

Copyright Notice © 2019 Real Time Automation, Inc. All rights reserved. Printed in USA.

This document is copyrighted by Real Time Automation Inc. Any reproduction and/or distribution without prior written consent from Real Time Automation, Inc. is strictly prohibited.

Trademark Notices
Allen-Bradley, ControlLogix, FactoryTalk, PLC-5, Rockwell Automation, Rockwell Software, RSLinx, RSView, are registered trademarks of Rockwell Automation, Inc.

ControlNet is a registered trademark of ODVA Inc.
DeviceNet is a trademark of the ODVA Inc.
EtherNet/IP is a trademark of the ODVA Inc
Modbus TCP is a trademark of Modbus IDA
Ethernet is a registered trademark of Digital Equipment Corporation, Intel, and Xerox Corporation.
Microsoft, Windows, Windows ME, Windows NT, Windows 2000, Windows Server 2003, and Windows XP are either registered trademarks or trademarks of Microsoft Corporation in the United States and/or other countries.

All other trademarks are the property of their respective holders and are hereby acknowledged

PACKAGING PRODUCT MANAGER'S CONNECTIVITY GUIDE

DEDICATION

To the Automation Engineer, the unsung hero of American Manufacturing

PACKAGING PRODUCT MANAGER'S CONNECTIVITY GUIDE

TABLE OF CONTENTS

DEDICATION 3
TABLE OF CONTENTS 4
INTRODUCTION 5
FIRST THINGS FIRST: ACTION PLANS 9
THE EIGHT TOP CUSTOMER REQUIREMENTS ... 17
THE TOP PACKAGING INDUSTRY NETWORKS . 24
FACTORY FLOOR SECURITY 43
INTEGRATING MACHINES WITH PACKML 46
IF YOU STILL USE ASCII… 51
GLOSSERY 54
ABOUT THE AUTHOR 60
OTHER BOOKS BY JOHN RINALDI 63

PACKAGING PRODUCT MANAGER'S CONNECTIVITY GUIDE

INTRODUCTION

It's been the best of times, and the worst of times, in the packaging industry lately. On one hand, we've seen incredible market growth over the last decade. The expansion of online shopping, the development of distribution centers around the country and the new-found brand awareness in the developing world continues to lead to significant growth. And unlike almost every other industry, packaging hasn't been adversely impacted (as yet) by digital technologies and the Internet. It's a time that arguably could be labeled a "boom time" and, the really good news, is that we can expect it to continue into the foreseeable future.

On the other hand, we know that what goes up can come down. There are significant headwinds that could lead to challenges in the future:

- Technologies (of all sorts) are coming online at a faster rate than our customers can understand, plan and absorb.

- New business models, new processes and paradigms are changing how our customers do business in ways we don't fully understand yet.

- Sustainability issues pose an unknown threat. What we do know is that sustainability is a growing concern to consumers around the world, and the amount of packaging that winds up in our oceans and landfills is a very real problem.

PACKAGING PRODUCT MANAGER'S CONNECTIVITY GUIDE

- Increasing customer requirements from the conversion of control systems based on traditional OT (Operational Technology) into systems that combine OT and IT. And, all those requirements are vastly expanding and changing the connectivity they expect (require) from packaging suppliers.

It's that last point that I address in this guide that I personally wrote for you. My name is John Rinaldi and I am the President and CEO of Real Time Automation and when I think about how difficult it is to sell automation systems to companies in pharmaceuticals, automotive, consumer products or food and beverage compared to just five years ago, my head spins.

Just a few short years ago, customers in all our prime markets would gladly accept a standalone packaging system. Today, it's not unusual to get requests for one or more types of controller communications technologies at the controller/actuator level: a different one for IT, MES or ERP integration; and another one for cloud communications. Sometimes they'll need data to archive in their chosen database, sometimes to feed an analytics system and sometimes to link into a proprietary IT system. (Sometimes the people we're talking to don't really know why they want something; it's just a check box on their requirements.)

The bad news is that it's going to get worse before it gets better. Predictive analytics is quickly being implemented by all our best Fortune 500 customers. Industry 4.0 connectivity and solutions that ride on top of architectures like OPC UA are more and more prevalent. And cybersecurity is expected to become a requirement for factory floor control systems. Is your head spinning yet?

It's both difficult and essential to have a team that can support these kinds of customer requirements, build those technologies – the right technologies – into your product, and support customers who need customization. Our packaging customers have found that having a partner like our company that can help you find your way through this morass and deliver the right solutions to your customers is both cost effective and speeds time to market. Our company specializes in delivering connectivity solutions to equipment providers in the packaging industry that are providing equipment and systems to manufacturing end users.

I wrote this short guide for my colleagues in the packaging

PACKAGING PRODUCT MANAGER'S CONNECTIVITY GUIDE

industry that are struggling to understand their manufacturing customers — what connectivity solutions they need, where digital manufacturing is headed and how best you can prepare your products for what lies ahead.

In this short guide, you'll find:

- A list of trends that are driving your manufacturing customers (think IT /OT conversion and Industry 4.0).

- A short, non-technical explanation of the most important connectivity standards that many of your customers are adopting.

- A glossary of terms that are specific to manufacturing connectivity that you should know.

- A special section that focuses on supporting those legacy systems that use old, ASCII Command/Response communications.

And, most importantly, an action plan that you can use to best position your company for future success in manufacturing. It's generic but appropriate for many suppliers in packaging. And because of its importance, it's the very next chapter.

Best Wishes for Success,

John

John Rinaldi
2019

PACKAGING PRODUCT MANAGER'S CONNECTIVITY GUIDE

Get More Rinaldi…

Whether you want to stay current with industry trends and technology, have a little bit of fun or qualify for some free swag, the **Best Darn Newsletter** from RTA serves it all up in one place.

And best of all; You won't have to delete it from your inbox. It gets handed to you personally by one of those brave men and women of the US Post Office. You know them; the ones that brave rain, snow, snarling dogs and feral cats to deliver the mail (printed text messages for you youngsters).

Sign up today at
https://www.rtautomation.com/company/newsletter/

> "This is my favorite newsletter! It has a little bit of everything: editorial, trivia, technology insight and I have a collection of the free giveaways at my disk."
>
> Karen Walkowiak
> Senior Project Control Engineer
> ASEA

PACKAGING PRODUCT MANAGER'S CONNECTIVITY GUIDE

FIRST THINGS FIRST: ACTION PLANS

In this chapter you're going to find a list of recommendations that are designed to assist you in meeting the complex and increasing requirements of manufacturing customers.

What follows is a series of recommendations that describe the technologies that are most likely to be important to your customers. It describes the media (how data moves from place to place) that you'll need to best serve your customers. It offers some cybersecurity recommendations to provide you with a few ideas on how to approach product development in this rapidly changing area. Implementation recommendations then provide some options ideas on how to go about supporting the connectivity requirements of these important customers. The chapter ends with some final recommendations.

> **DISCLAIMER**: As with most things, a few words of disclaimer are necessary. This book is written for the general class of packaging suppliers, it can't possibly describe the perfect plan for your organization, your product line, your market and your customers. Getting that specific requires some detailed discussions with your product planning and development staff. What follows here is simply some general recommendations that may not be appropriate to your particular situation.

PACKAGING PRODUCT MANAGER'S CONNECTIVITY GUIDE

TECHNOLOGY RECOMMENDATIONS

As a Product manager in the packaging industry, there are technologies you can't afford to miss and those you can do without . Technologies you should support include:

- EtherNet/IP – EtherNet/IP is the key to the kingdom that is known as the Rockwell Automation ecosystem. It is difficult to find success providing any product in the North American manufacturing market without supporting EtherNet/IP.

- PROFINET IO – Just as EtherNet/IP is the key to the Rockwell Automation architecture, PROFINET IO is the key to the Siemens eco system. PROFINET IO provides you with access to some important north American customers and is mandatory for access to the European market.

- Modbus TCP – There is little to lose by supporting this Ethernet protocol. Modbus TCP offers you a generic mechanism for moving data in and out of your device at little to no cost.

- OPC UA – OPC UA is an architecture for industrial connectivity that is growing in support. It offers mechanisms for moving data beyond the factory floor and is useful for connecting factory floor devices to middleware, IT applications and the cloud. Unlike most other technologies in this list, OPC UA offers a buffet of options and you should discuss your approach to OPC UA with someone who thoroughly understands the technology.

- HTTP – HTTP is a standard way for computers to connect. It is how browsers connect to servers. Supporting HTTP with a data formatting mechanism like XML or JSON provides you with a standard way to send important status information to a production system, maintenance system or other IT type application that your customer may be using.

- MQTT – MQTT is a popular, though less cybersecure mechanism, for transferring data from a device to the cloud

PACKAGING PRODUCT MANAGER'S CONNECTIVITY GUIDE

through an MQTT proxy device. It is designed for repeatedly transferring small bits of information.

- PROPRIETARY – Industry Heavyweights like SAP or OSI PI have custom and/or proprietary protocols that you may need to support depending on your customer requiremetns.

Technologies that are unlikely that you'll need and less important for you to support include:

- EtherCAT – EtherCAT enables high speed communications and is often found in motion control applications. EtherCAT is extremely popular for high speed applications.

- IO-Link – IO-Link is a low level, point-to-point IO system for easily connecting sensors and actuators to programmable controllers. Not as important for more systems that process data instead of I/O.

MEDIA RECOMMENDATIONS

A communications media is the physical medium that is used to move signals from your device to a destination. Ethernet cabling is a media, as is USB, serial and air (wireless). Ethernet is, of course, the most important media to support. Wireless is growing in popularity and will likely become the most important technology of the future. Wireless Ethernet is the simplest to support while the mesh technologies (ISA100, WirelessHart, Zigbee and Bluetooth) require more complicated configuration and interface devices.

Many of the Ethernet application layer protocols support some sort of ring technology. It is highly recommended that you provide two RS45 jacks on your device and the capability to support ring and linear (daisy-chained) Ethernet topologies.

There is no easy answer regarding wireless technology other than wireless Ethernet. For short distance communication, Bluetooth seems to be the clear winner. For more complicated networks, where nodes are dispersed, any of the wireless mesh technologies can be a good choice. There are long complicated arguments why you might choose Zigbee, Z-Wave, Thread or one of the other technologies but those won't be repeated here.

PACKAGING PRODUCT MANAGER'S CONNECTIVITY GUIDE

CYBERSECURITY RECOMMENDATIONS

Cybersecurity is an important topic. More and more time and attention is being devoted to protecting your customers factory infrastructure than at any time in the past. Unfortunately, only a few limited standards have emerged and there is no consensus on how best to protect the factory floor.

CIP Security for EtherNet/IP is the only technology that has emerged and seems to be ready for test deployments in the 2020. If you support or plan to support EtherNet/IP, CIP Security needs to be on your development roadmap.

ISA-62443 is another important standard. It is an emerging standard and it's not entirely clear if the market is going to embrace it or not.

Unfortunately, cybersecurity is such a new and rapidly changing area, there are only a few recommendations that can be made:

1. Invest in secure boot technology. Secure boot technology ensures that the software in your device is the software provided by the original device manufacturer.

2. Make plans to protect the private keys of your customers PKI (Private Key Infrastructure) in hardware. It is likely that factory floor devices will be moving toward certificates and private keys and there is no secure mechanism to protect keys in software.

IMPLEMENTATION CONSIDERATIONS

How to get the connectivity you need for your device? Manufacturing customers have wide ranging requirements and use many different connectivity solutions. There are many options to consider as you plan to meet these expansive requirements and many different costs to consider. Your first decision is to get the connectivity from either software or hardware. There are advantages and disadvantages to both.

Hardware connectivity solutions offer:

- A dedicated communications coprocessor that offloads processing of protocols from your device. This can be extremely important if your device already contains

PACKAGING PRODUCT MANAGER'S CONNECTIVITY GUIDE

complicated or certified software. Avoiding the adding of complicated software to an already complicated device application can save significantly more effort and testing as well as the possibility of unexpected interactions with the communication software.

- Faster time to market – generally you can deliver more connectivity faster with a hardware solution than with implementing connectivity in software

- A 3rd party to continually update and support the off-the-shelf system freeing your staff to work on the features of your device and not connectivity

- More connectivity at lower cost if you find a supplier with a connectivity solution that offers multiple protocol solutions

The one big disadvantage to hardware connectivity – BOM cost. There are a series of considerations to keep in mind when selecting a hardware solution:

HOST API – How will your software application communicate with the connectivity PCB? Some solutions in the market use very difficult structure and are take notoriously long efforts to implement. This sometimes destroys the ability to get to market quickly.

FOOTPRINT – The size of the board is important as space is needed to provide for it. The biggest consideration is how the physical ethernet connection will be made from the customers network to the connectivity PCB.

POWER – What type and how much power does the PCB require?

INDICATORS – LED indicators are one of the things that are often overlooked. Some of the technologies described earlier can use indicators if they are available.

PACKAGING PRODUCT MANAGER'S CONNECTIVITY GUIDE

Software connectivity solutions offer:

- No BOM costs

- More highly tailored, customized solutions that take the specifics of your device into consideration

- The option of using open source solutions

The one big disadvantage of software connectivity solutions is the time, labor and ongoing support required from your engineering staff. Those costs can be significant. Technologies like PROFINET IO revise their specification every 24 months require users to conform to the revised specification. Suppliers can spend 3 to 6 months installing the latest version. Many product managers don't realize how expensive software connectivity can be and minimize these costs when beginning a connectivity project.

There are several things to consider when looking at these solutions:

- No special hardware required - Most, but not all, of the technology solutions described in this guide can be implemented in software. No special hardware other than an Ethernet port is required. Technologies like EtherNet/IP, PROFINET IO, OPC UA and Modbus TCP are, for the most part, application layers that are embedded in the standard TCP or UDP protocols.

- An operating system or a task scheduler is usually required - Any of these technologies can be implemented as tasks within any software system that uses an operating system like Linux, MS Windows 10, VxWorks or other embedded operating system. It can be as simple as creating a new task that can manage communication over Ethernet using one of those industrial Ethernet application layers.

- Data Modeling is the key consideration – Software implementation is not the most difficult implementation consideration. What is more difficult is the assignment of your operating data to the specific structures defined for the

PACKAGING PRODUCT MANAGER'S CONNECTIVITY GUIDE

technology you are implementing. For Modbus TCP, that means assigning your data to unsigned integer Registers and binary Coils. For OPC UA, that means grouping your data into objects, creating nodes and assigning your data to variable nodes. For PROFINET IO, it means grouping your data into a Rack/Slot/Module/Point structure. This is the key implementation consideration because this is the structure that users will use when they access your device from a controller or a Windows or Linux tool. It is not a task that should be taken on lightly.

Another key implementation consideration is where you source your connectivity software. You have three possibilities:

Build it Yourself – Some companies insist on owning everything building everything from the ground up. If you company has the time, money and expertise to design and build connectivity software, it can be done. Most companies have better things for the engineers to do and choose to get software that isn't core to their processes or key value. Building it yourself does not provide you with any additional insight into how best to configure how end users will access your device.

Open Source – Open source solutions in industrial automation are now more prevalent than ever. These solutions allow device developers to incorporate solutions at what appears to be no cost. Unfortunately, with uncertain maintenance, no assistance in creating an object model, no integration assistance and no certainty that the open source software will be maintained as the specification evolves, open source can be the most costly solution you may choose.

Purchased solutions – Vendors like Real Time Automation make software solutions available. These solutions come with integration assistance, maintenance, object model design assistance and training for your sales and support staff.

The considerations outlined in this chapter are general in nature. Your specific application, your customers specific requirements and your competitive environment must be considered when forming a comprehensive connectivity plan for your device.

PACKAGING PRODUCT MANAGER'S CONNECTIVITY GUIDE

<u>Help is available…</u>

Connectivity is a complicated area and it is a mistake to not use experts to get it done right. You can get that expertise from companies like Real Time Automation. Our company (and others) offer expert assistance in analyzing your specific requirements and forming a detailed connectivity plans for your device.

Get a free technology assessment from the connectivity experts at RTA.

You can schedule that by calling 262-436-9299, emailing <u>sales@rtautomation.com</u> or visiting <u>https://www.rtautomation.com/contact/</u>.

THE EIGHT TOP CUSTOMER REQUIREMENTS

The world is undergoing a metamorphosis of technological, cultural and societal shifts that dwarf any other period in world history. Harnessing water power, the steam engine, the assembly line, even the birth of the PLC, do not compare with the pace or scope of changes we've seen so far in the 21st century. And there is no sign that it will slow down enough for us to catch our breath.

In manufacturing, naturally risk-adverse and capital-intensive customers slow the pace of change. But even here, we've seen an explosion of new technologies, operating theories and business models This makes it exceedingly difficult to plan for product enhancements in the future. You need to take these factors (listed in priority order) into consideration as your product plans are developed.

1. MORE SMART SELF-MANAGED PRODUCTS

Manufacturers in North America are confronting the most difficult staffing environment of the last 50 years. A continuing drop in the birth rate, a steadily increasing rate of retirement by experienced, baby-boomer staff, a decrease of people interested in manufacturing as a career, and an increasingly tight job market are creating a nightmare scenario for manufacturers.

Analysis: The staff reduction in manufacturing plants dictates a compete rethinking of how products are configured, how they operate, and how error/failure conditions are managed. Automation product vendors must create products that are self-configurable, secure and operable without user interaction. These products will learn, make decisions, self-diagnose and heal themselves.

2. SPLITTING OF THE AUTOMATION ENVIRONMENT

There is a growing evidence that the automation environment will split into two distinct segments – control and information. Control will be the traditional control offered by programmable controllers using Ethernet and I/O systems for moving real time I/O data. Everything else will be information products that look more like IT products than they do today. Organizations are likely to split along the same lines. Control systems will be governed by a control's group while everything else is likely to be managed by a segment of IT dedicated to supporting manufacturing. Distributors are now angling to fill this role by creating IT/OT teams that can support the traditional real-time manufacturing staff and the IT side of the production floor.

Analysis: Product managers must know which side of automation environment they are on and position their products for that market. A valve controller must be marketed differently than a historian. The product features, customer types, sales channels and pricing may all vary depending on the segment chosen.

3. MORE EMPHASIS ON CYBERSECURITY

Ethernet has brought a vast productivity improvement to the factory floor, but it's also enabled cyberattacks that were nearly impossible in the more closed architectures of yesterday. Hackers have shifted their attention from MS Windows system to the more enjoyable and lucrative world of factory automation. Because the majority of these attacks are not publicized, no one knows for certain how many plants have had servers locked, important data stolen, messages altered and controllers hijacked. Management has recognized the great risk to factory floor cyberattacks and efforts are underway to secure to factory floor networks. Unfortunately, it isn't clear what standards may be adopted, what practices will be put in place, or how many customers will implement cybersecurity.

Analysis: There is no question that cybersecurity is an important concern. What will be required of automation products in the future is anyone's guess. Product managers will need to monitor developments in cybersecurity, learn from customers how they plan to manage this risk, and adapt their products.

4. MORE WIRELESS MANUFACTURING

Consumer society is now wireless. Just as the factory floor followed the larger (consumer) society in the deployment of Ethernet, it is following the deployment of wireless connectivity (albeit at a much slower pace). More wireless Ethernet (802.11), 5G communications, improved 4G LTE, Bluetooth and spread spectrum technologies are destined to make the factory floor more wireless than wired in a few short years. The newest generation of plant engineers and managers is more comfortable with deploying wireless communications than their predecessors, and more anxious to reduce costs, increase connectivity and capture more and more data.

Analysis: Equipping your automation product for wireless communication must be on your product development road map. Depending on your product category, it might be 802.11, 4G, Bluetooth or a spread spectrum solution. There is no question that 5G will come to the factory floor and, if it follows recent technology trends, it will probably happen sooner than we expect.

5. MORE DATA AND MORE OPEN COMMUNICATION

Data sources, the quantity of data, how data is organized and moves around the factory floor is undergoing massive change. Manufacturers are sensorizing the factory floor like never before. Pressure, temperature, stress, strain, vibration and hundreds of other variables are being captured for the first time. More manufacturers are expecting every machine component to yield operational, quality and diagnostic data. And all that data is more often than not required to be transferred around the factory by more open "IT-like" protocols and open technologies using open standards, so data easily flows to where it is needed.

Analysis: Automation products are going to be expected to deliver more process data, maintenance data and analytical information in the near future. Operational data will be required to be delivered to PLCs using hard real time communication protocols. Non-operational data, such as quality, maintenance and archival, must be delivered using open standards like JSON/XML over HTTP and OPC UA, with standard data models that simplify data digestion. Automation products must become more IT-like in features, operating characteristics and network interfaces. As controls departments become more IT-oriented, vendors supplying factory floor automation must create products for people with IT skills.

6. MORE INTENSE FOCUS ON INTEGRATION COSTS

Nothing is more costly to a manufacturer than integration of, a component, machine or system. Whether it is simply adding a linear actuator or integrating a new packaging system, the time from arrival of the component on the floor until it is producing product is extremely expensive. Customers will standardize on architectures, protocols and object models to lower these costs.

OPC UA is a leading technology that can lower these integration costs. With OPC UA, these machine components use common security, encodings and transports. Combining OPC UA with a data modeling standard like PackML solves this integration problem for manufacturers.

Analysis: OPC UA with PackML meets the need for faster, easier machine integration. OPC UA is already a popular technology with manufacturers and product managers should have it on their road map. PackML, the packaging machine standard, may also be adopted as the mechanism for enforcing a consistent data model on top of OPC UA. Product managers should watch these developments closely and be ready to add PackML (or an equivalent) if it becomes the favorite of the big manufacturing companies.

# PACKAGING PRODUCT MANAGER'S CONNECTIVITY GUIDE

7. COMMODITIZATION OF HARDWARE (THE RASPBERRY PI PHENOMENA)

Microsoft's Satya Nadella is famous for saying in 2015 that every business is a software business. Some in manufacturing may have mocked him for that, but it's becoming true. Case in point: the newest Raspberry PI, the Raspberry PI 4 B. For the first time, the Raspberry PI is powerful enough to be used as a full-fledged personal computer. This product is a giant step forward in low-cost, highly functional embedded computing. It is loaded with features, functionality and processor bandwidth that surpasses the power of much of what we find on the factory floor.

Analysis: Hardware development is now commoditized. With off-the-shelf modules that can be incorporated into an industrial product for as little as $35, we are at the point where its uneconomical to build your product around fully custom hardware. Product managers must wisely evaluate the time-to-market, and what part of product content should be owned and what part should be off-the-shelf.

8. DEMAND FOR SPEED AND EASE OF USE

The younger members of society are a product of an on-demand world that has always delivered what they want, when they want it and how they want it. Pizzas, movies, rides, books are delivered nearly instantly. As this generation begins to populate today's manufacturing facilities, they are going to bring those expectations to your products and company. They will expect a next-generation speed and ease of use far beyond the requirements of older generations of customers.

Analysis: Automation product managers must engineer products with the expectations of the younger engineers and managers. Product, data, support, information delivery must be streamlined to meet the requirements of these new customers.

PACKAGING PRODUCT MANAGER'S CONNECTIVITY GUIDE

Like What You See?

If you're enjoying the book and still have a jones for more…

…make sure you pop over to John's blog at https://www.rtautomation.com/enginerd-exclusive/ and sign up for email alerts on new posts. [There's a library worth of archives on the site, too.]

Our promise? You'll learn something and be entertained!

THE TOP PACKAGING INDUSTRY NETWORKS

Networking technologies continue to evolve but there are a number of them that will always be found in North American manufacturing plants. On the next few pages, you'll find these technologies.

PACKAGING PRODUCT MANAGER'S CONNECTIVITY GUIDE

ETHERNET/IP™

Why It's Important to Product Managers

EtherNet/IP provides the keys to the kingdom that is the ecosystem of Rockwell Automation and the Allen-Bradley brand. With EtherNet/IP you have access to 60-70% of the automation market in North America.

Quick Overview

Ethernet/IP is an implementation of CIP, the Common Industrial Protocol. CIP organizes and shares data in a way that is completely independent of Transport, Media Access and Physical Layer communications. CIP is the core for DeviceNet, ControlNet, CompoNet and EtherNet/IP. It provides EtherNet/IP with basic messaging and structure for managing data as objects.

EtherNet/IP is the Ethernet implementation of CIP. It provides the following:

Common data addressing - An object-based structure inherited from CIP which organizes device data as a set of objects with attributes. There are a set of required objects which are identical for all EtherNet/IP devices and a set of objects that are specific to the device application.

I/O exchange – I/O is exchanged between a controller (termed a Scanner) and an end device (termed an Adapter). Inputs are cyclically delivered to the scanner from the Adapter over UDP (User Datagram Protocol) while outputs are cyclically transferred from the scanner to the adapter.

Data exchange – a mechanism for scanners and EtherNet/IP tools to deliver configuration and operational parameters to an adapter.

Security - A secure version of EtherNet/IP is being introduced in 2020. This standard requires authentication and integrity of EtherNet/IP messages. It compels both Scanners and Adapters to authenticate to ensure that they are the devices they claim to be and message encryption to ensure accuracy and privacy. Using secure EtherNet/IP confidential communications between trusted devices are allowed and communication between untrusted entities are disallowed.

EtherNet/IP is the mechanism that Rockwell Automation ControlLogix processors use to exchange data with Ethernet devices.

Trade Organization

The ODVA in Ann Arbor, Michigan. (https://www.odva.org/)

Certification

Vendors offering EtherNet/IP devices must register with the ODVA and certify their devices in the ODVA test laboratory.

How to Get Started?

Either hardware or software mechanisms can be used to enable a device for EtherNet/IP. Call RTA 262-436-9299 and speak to an application engineer, email sales@rtautomation.com or visit https://www.rtautomation.com/contact/ to learn more.

Resources

Get the EtherNet/IP book to learn more about EtherNet/IP and how to implement it in your device. Visit https://www.rtautomation.com/product/ethernet-ip-the-everymans-guide-to-ethernet-ip/ for more details.

PACKAGING PRODUCT MANAGER'S CONNECTIVITY GUIDE

PROFINET

Why It's Important to Product Managers

PROFINET IO is the Ethernet technology communication standard used by Siemens Programmable Controllers. As Siemens is the dominant controller provider in Europe, PROFINET IO is the key technology that enables access to manufacturing systems throughout most of Europe.

Quick Overview

PROFINET IO is a high-level, Ethernet network for industrial automation applications. Built on standard Ethernet, PROFINET IO uses traditional Ethernet hardware and software to define a network that structures the task of exchanging data, alarms and diagnostics with Programmable Controllers and other automation controllers. PROFINET IO uses cyclic data transfer to exchange data with Programmable Controllers over Ethernet. A Programmable Controller and a device must both have a prior understanding of the data structure and meaning.

PROFINET IO provides the following:

Device classification – a method of organizing devices where devices are organized into one of three types; IO-Controllers, IO-Devices and IO-Supervisors. IO-Controllers are devices that execute an automation program. Controllers exchange data with IO-Devices. IO-Devices are distributed sensor/actuator devices connected to the IO-Controller over Ethernet. IO-Supervisors are HMIs, PCs or other commissioning, monitoring or diagnostic analysis devices..

Common data addressing – PROFINET IO organizes a network where every device is accessed as a remote I/O rack with slots, modules, channels and points. Vendors supplying devices provide a control file that identifies how a device is organized and what I/O data it contains. The control file contains the device data needed by a Profinet IO Controller to access the end-device.

I/O exchange – I/O is exchanged between an IO-Controller and an IO-Device. Inputs and outputs are cyclically transferred between the IO-Controller and the IO-Device over UDP (User Datagram Protocol).

PACKAGING PRODUCT MANAGER'S CONNECTIVITY GUIDE

PROFINET IO is the mechanism that Siemens S7 Controllers use to exchange data with Ethernet devices.

Trade Organization

PI North America in Phoenix, Arizona (https://us.profinet.com/about/pi-north-america/)

Certification

Vendors offering PROFINET IO devices must register with PI North America and certify their devices in the PI North America test laboratory.

How to Get Started?

Either hardware or software mechanisms can be used to enable a device for PROFINET IO. Call RTA 262-436-9299 and speak to an application engineer, email sales@rtautomation.com or visit https://www.rtautomation.com/contact/ to learn more.

Resources

Call RTA 262-436-9299, email sales@rtautomation.com or visit https://www.rtautomation.com/contact/ to get the white paper describing PROFINET IO.

PACKAGING PRODUCT MANAGER'S CONNECTIVITY GUIDE

MODBUS TCP

Why It's Important to Product Managers

Modbus TCP, the successor to serial Modbus RTU, is important today as there are thousands of devices that use this simple mechanism for moving manufacturing data from simple devices to PCs and control systems. Its simplicity makes it a popular choice for many device vendors.

Quick Overview

Modbus is a method used for transmitting information over serial lines between electronic devices. Originally intended for communications between programmable logic controllers (PLCs) and computers in the days before Ethernet, it was the de facto standard communication protocol for connecting a wide range of industrial electronic devices.

Modbus TCP is the successor to serial Modbus and simply adapts the serial communication protocol for Ethernet. The Modbus protocol is wrapped in an Ethernet TCP packet. Devices which support serial Modbus can send and receive Modbus TCP packets that are nearly identical.

Like serial Modbus, Modbus TCP's raw simplicity gives it a number of distinct advantages:

Simple Data Representation – Modbus TCP has only two basic data types, a 16-bit unsigned integer (known as a register) and a single bit (known as a coil).

Simple Request/Response Command Structure – Modbus TCP has a simple read and write command for each of the different data types.

Low Resource Requirements – Modbus TCP requires very little in the way of processor code space or RAM. This isn't as important today given the powerful processors and technology available to us, but it was very important in the early years of industrial automation when processors used 8-bit technology and resources like RAM and ROM were extremely expensive and scarce.

Ethernet TCP Transport Layer – Modbus TCP uses TCP (Transmission Control Protocol) to move data from one place to another. This means that any processor can implement Modbus TCP if it supports Ethernet.

Trade Organization

The Modbus Organization in Hopkinton, Massachusetts (http://www.modbus.org/)

Certification

Unlike many other standards, Modbus TCP certification is optional. Users can self-certify their products with the Modbus TCP Conformance Test Program available at the Modbus Organization web site.

How to Get Started?

All that is required to enable Modbus TCP is a working Ethernet device. Software to build Modbus TCP applications is freely available at a number of open source sites and available from RTA. Call RTA 262-436-9299 and speak to an application engineer, email sales@rtautomation.com or visit https://www.rtautomation.com/contact/ with questions.

Resources

Get the Modbus book to learn more about serial Modbus and Modbus TCP and how to implement it in your device. Visit https://www.rtautomation.com/product/modbus-the-everymans-guide-to-modbus/ for more details.

PACKAGING PRODUCT MANAGER'S CONNECTIVITY GUIDE

ETHERCAT

Why It's Important to Product Managers

EtherCAT is one of the fastest technologies for moving machine data around the factory floor. Product managers with automation products that can leverage faster response times should invest in EtherCAT technology.

Quick Overview

EtherCAT is a highly flexible Ethernet network that uses something called "processing on the fly." The leading byte of an EtherCAT byte stream is passed to the next node in the ring while it is being processed by the current node, providing the network with blazing speed and efficiency.

Unlike most other technologies, a single message is issued by the EtherCAT Master with data for all nodes. As the message is transmitted around the EtherCAT ring and back toward the Master, each node reads its inputs and adds its outputs to the message. When the message arrives back at the EtherCAT Master, every node in the network has received new input data from the Master and returned new output data to the Master. Without the deficiencies of small payloads or messages targeted to specific nodes, an EtherCAT network can achieve maximum bandwidth utilization.

EtherCAT is a complete solution. It includes, among other things, a safety protocol and multiple device profiles. EtherCAT also benefits from a strong user group. It offers a strong combination of benefits:

Higher bandwidth utilization than any other competitive technology. Typical EtherCAT cycle times are 50…250 μs, while traditional fieldbus systems take 5-15 ms for an update.

EtherCAT is very precise. It distributes clocks around the network and that set of distributed clocks provides measurement values that can be sampled and allows for setting outputs in a synchronized manner network-wide. Jitter is substantially less than one microsecond.

PACKAGING PRODUCT MANAGER'S CONNECTIVITY GUIDE

The topology is extremely flexible. There are no practical limitations regarding topology – line, star, tree, redundant ring and all those combined with up to 65535 nodes per segment.

EtherCAT is inexpensive though hardware is required. Slave controller chips are available and several different microprocessors embed EtherCAT technology.

Safety and control are combined on the same EtherCAT network. The safety protocol is based on the application layer of EtherCAT, without influencing the lower layers. It is certified according to IEC 61508 and meets requirements of Safety Integrity Level (SIL) 3.

Trade Organization

The EtherCAT Technology Group (ETC) in Carlsbad, CA. (https://www.ethercat.org/default.htm)

Certification

EtherCAT products must pass an in-house test with the Conformance Test Tool (CTT) according to the Conformance Test Policy and the ETG Vendor ID Agreement.

How to Get Started?

Hardware is required to enable a device for EtherCAT. Call RTA 262-436-9299 and speak to an application engineer, email sales@rtautomation.com or visit https://www.rtautomation.com/contact/ to learn more.

Resources

The ETC periodically offers training classes for end users at its Carlsbad office in California. See the ethercat.org website for more details.

Questions?

PACKAGING PRODUCT MANAGER'S CONNECTIVITY GUIDE

OPC UA

Why It's Important to Product Managers

OPC UA is the next generation of the OPC technology that's been deployed over the last twenty-five years. It provides additional open transports, better security, and a more complete information model than traditional OPC. UA provides a flexible and adaptable mechanism for moving data between enterprise type systems and the kinds of controls, monitoring devices, and sensors that interact with real world data. UA is the best technology for connecting databases, analytic tools, and other enterprise systems with real world data from controllers, sensors, actuators, and monitoring devices that interact with real processes that control and generate real world data.

Quick Overview

OPC UA, like its factory floor cousins (EtherNet/IP and PROFINET IO), is composed of a client and a server. The client device requests information. The server device provides it. But what the UA server does is much more sophisticated than what EtherNet/IP, Modbus TCP or ProfiNet IO servers do.

An OPC UA Server models data, information, processes, and systems as Objects and presents those Objects to Clients in ways that are useful to vastly different types of Client applications. And better yet, the OPC UA Server provides sophisticated services that the Client can use including:

Discovery Services – services that Clients can use to know what Objects are available, how they are linked to other Objects, what kind of data and data type is available, what meta-data is available that can be used to organize, classify, and describe those objects and values.

Subscription Services – services that the Clients can use to identify what kind of data is available for notifications. Services that Clients can use to decide how little, how much and when they wish to be notified about changes.

Query Services – services that deliver bulk data to a Client, like historical data for a data value.

Node Services – services that Clients can use to create, delete, and modify the structure of the data maintained by the Server.

Method Services – services that the Clients can use to make function calls associated with Objects.

Unlike the standard industrial protocols, an OPC UA Server is a data engine that gathers information and presents it in ways that are useful to various types of OPC UA Client devices. Those devices could be located on the factory floor like an HMI, a proprietary control program like a recipe manager or a database, dashboard, or a sophisticated analytics program that might be located on an enterprise Sever.

Trade Organization

OPC Foundation in Phoenix, Arizona (https://opcfoundation.org/)

Certification

Vendors offering OPC UA devices must register with the OPC Foundation and certify their devices through an accredited test laboratory.

How to Get Started?

Either hardware or software mechanisms can be used to enable a device for OPC UA. Call RTA 262-436-9299 and speak to an application engineer, email sales@rtautomation.com or visit https://www.rtautomation.com/contact/ to learn more.

Resources

Get the OPC UA book to learn more about OPC UA and how to implement it in your device. Visit https://www.rtautomation.com/product/opc-ua-unified-architecture-the-everymans-guide-to-opc-ua/

PACKAGING PRODUCT MANAGER'S CONNECTIVITY GUIDE

IO-LINK

Why It's Important to Product Managers

IO-Link is one of the most popular technologies for moving sensor and actuator data into a controller. It is a point-to-point communication protocol that is generally not appropriate for more complicated devices like motor drives, printers, inspection equipment and other devices that have significant quantities of data.

Quick Overview

IO-Link is a master / slave IO network with point-to-point connections between a master and a set of devices. An IO-Link system consists of an IO-Link Master, one or more IO-Link Devices, cabling from each device to the master and an Engineering tool to configure and assign parameters.

The IO-Link Master manages the connection with each of the IO-Link Devices. It can be installed either in the control cabinet or in the field. An IO-Link Master contains a limited number of IO-Link ports (8 typical) each of which communicates to an IO-Link Device. Simple 3-wire cabling (up to 20 m long) is used to connect each port on the master to quick connect M5, M8 or M12 connectors on the device. Typically, the IO-Link Master has an Ethernet connection into an automation system using an Ethernet protocol like EtherNet/IP, PROFINET IO or Modbus TCP.

The master's IO-Link ports can be operated in one of four modes. In "IO-Link" mode, the port is used for IO-Link communication. In "DI" mode, the port functions as a digital input. In "DQ" mode, the port functions as a digital output. Lastly, in "Deactivated" mode, the port is disabled.

IO-Link Masters are simple sensor and actuators that collect inputs and set outputs. The connection from IO-Link Master to IO-Link Device is over a cable that is a simple point-to-point communication link and not a fieldbus. Connections are configured using the IO-Link Engineering tool.

IO-Link offers a strong combination of benefits including:
Easy wiring
Automated parameter setting

Extended diagnosis

Trade Organization

The IO-Link Consortium in Karlsruhe, Germany (https://io-link.com/en/)

Certification

There is a standard test specification for IO-Link Master and devices. The specification is maintained by the IO-Link consortium and provides the basis for device tests that are done at the IO-Link competency center.

How to Get Started?

Hardware is required to enable a device for IO-Link. Call RTA 262-436-9299 and speak to an application engineer, email sales@rtautomation.com or visit https://www.rtautomation.com/contact/ to learn more.

Resources

The IO-Link consortium periodically offers workshops to train vendors and end users. See the consortium website for more details.

Questions?

Contact the author at jrinaldi@rtautomation.com or call 262-436-9299.

PACKAGING PRODUCT MANAGER'S CONNECTIVITY GUIDE

HTTP

Why It's Important to Product Managers

HTTP is the main protocol of the Internet. All browsers communicate using HTTP and its secure version, HTTPS. Supporting HTTP in an industrial product provides connectivity to many middleware and IT type applications.

Quick Overview

HTTP is a very simple technology and many vendors have built applications on top of it to move data between automation devices and IT applications. Automation product vendors can create easy to implement mechanisms for moving automation data between IT systems and the factory floor using HTTP.

HTTP is a request / response protocol that operates at a very low level. In its most often-used implementation on the Internet, the HTTP GET service is used to request a web page from a remote Server. The response message contains the HyperText Markup Language (HTML) that forms the web page content that appears in your browser.

Unlike many other computer protocols, HTTP closes the connection when a request completes. New connections are opened on every data transmission and that overhead makes it unattractive for factory I/O systems but acceptable for applications that move data between an OT (Operational Technology) device and an IT application.

HTTP, unlike some other IoT protocols, provides no information model, no services other than the raw GET and PUT, and no standardized mechanism to describe data. A GET request can return data and a PUT request can send data but there is no standard that defines what data is contained in the message. Some HTTP implementations use a proprietary body to encode data. Others use the well-known data markup languages JSON and XML.

The lack of a mechanism for describing commands and responses is a deficiency of HTTP. Despite this limitation, HTTP's popularity and simplicity make it a very popular protocol for IoT applications where building a proprietary implementation is not always undesirable.

Trade Organization

There is no trade organization for HTTP. HTTP is a protocol standard from the IETF (Internet Engineering Task Force) and part of all TCP/IP stacks.

Certification

No certification is available for HTTP applications or devices

How to Get Started?

HTTP is purely a software protocol. Any Ethernet device can support the HTTP transport layer, but a mechanism to describe the application data is required; either some proprietary or standard open description like a JSON or XML. To discuss what is best for your application, contact an RTA application engineer on 262-436-9299, email sales@rtautomation.com or visit https://www.rtautomation.com/contact/ to learn more.

Resources

There are many resources for HTTP and HTTPS on the world wide web.

PACKAGING PRODUCT MANAGER'S CONNECTIVITY GUIDE

MODBUS RTU

Why It's Important to Product Managers

Modbus RTU remains important today because of the thousands of legacy Modbus applications in factory installations all over the world. Though it has a large installed base, it's rare to find vendors delivering new automation applications with Modbus.

Quick Overview

Modbus is a method used for transmitting information over serial lines between electronic devices. Originally intended for communications between programmable logic controllers (PLCs) and computers in the days before Ethernet, it was the de facto standard communication protocol for connecting a wide range of industrial electronic devices.

Modbus has found its way into hundreds of thousands—if not millions—of devices. You can find it in everything from valve controllers, to motor drives, to HMIs, to water filtration systems. It would be difficult indeed to name a product category in Industrial or Building Automation that doesn't use Modbus. Its raw simplicity gives it a number of distinct advantages:

Simple Data Representation – Modbus has only two basic data types, 16-bit unsigned integer (known as a register) and a single bit (known as a coil).

Simple Request/Response Command Structure – Modbus has a simple read and write for each of its different data types.

Low Resource Requirements – Modbus requires very little in the way of processor code space or RAM. This isn't as important today given the powerful processors and technology available to us, but it was very important in the early years of industrial automation when processors used 8-bit technology and resources like RAM and ROM were extremely expensive and scarce.

Serial Transport Layer – Modbus uses RS485 serial communications to move bits from one place to another. This means that any processor can implement Modbus without any special hardware. All you need is a simple and inexpensive RS485 driver chip to be in the Modbus business.

Message Checking – CRC and LRC checking mean that transmission errors are checked to 99% accuracy.

Trade Organization

The Modbus Organization in Hopkinton, Massachusetts (http://www.modbus.org/)

Certification

Unlike many other standards, Modbus certification is entirely optional. Users can self-certify their products with the Modbus Conformance Test Program available at the Modbus Organization web site.

How to Get Started?

All that is required to enable Modbus is a UART (Universal Asynchronous Receiver Transmit) port and an RS485 circuit driver. The software is freely available at a number of open source sites. Call RTA 262-436-9299 and speak to an application engineer, email sales@rtautomation.com or visit https://www.rtautomation.com/contact/ with questions.

Resources

Get the Modbus book to learn more about Modbus and how to implement it in your device. Visit https://www.rtautomation.com/product/modbus-the-everymans-guide-to-modbus/ for more details.

PACKAGING PRODUCT MANAGER'S CONNECTIVITY GUIDE

MQTT

Why It's Important to Product Managers

MQTT is popular for sending data from automation devices to the cloud. Vendors with small data sets often select MQTT for moving that data to the cloud.

Quick Overview

MQTT is designed to meet the challenge of publishing small pieces of data in volume from many devices over a network constrained by low bandwidth, high latency, or uncertain reliability. MQTT is very useful in applications where a small set of devices are sending small data packages to proprietary servers.

The heart and soul of MQTT is it's publish/subscribe architecture. This architecture allows a message to be published once and go to multiple consumers, with complete message decoupling between the producer of the data/events and the consumer(s) of those messages and events. Data is organized by topic in a hierarchy with as many levels of subtopics as needed. Consumers can subscribe to a topic or a subtopic. They can also use wildcards to specify topics of interest.

An ASCII name is used to identify topics. A broker receives the information from the servers and matches the information to consumers by topic name. The information is distributed to every consumer that matches the topic. If no consumer requires the information, the information is discarded.

Some of the benefits of using MQTT in these applications include:

Very efficient event handling. MQTT is a "PUSH" system in which the producers push data to brokers. No bandwidth is consumed by consumers requesting data.

Low latency. Information is pushed immediately to consumers.

Low resources required by publishers. This makes it very good for low resource devices like sensors and actuators.

Very reliable on fragile and unreliable networks. Brokers can be configured to retain messages for consumers that are temporarily disconnected.

PACKAGING PRODUCT MANAGER'S CONNECTIVITY GUIDE

Side Note: MQTT is an insecure mechanism for sending data to the cloud. It should only be used in applications where privacy and confidentiality are not required.

Trade Organization

There is no trade organization for MQTT but there is an MQTT community where users share information and work together on MQTT issues. You can find that at http://mqtt.org/.

Certification

No certification is available for MQTT applications or devices.

How to Get Started?

MQTT is purely a software mechanism. Any Ethernet device can support the MQTT transport layer but an application layer must be used to enable connectivity. For more information about what is best for your application, contact an application engineer at RTA 262-436-9299, email sales@rtautomation.com or visit https://www.rtautomation.com/contact/ to learn more.

Resources

http://mqtt.org/

FACTORY FLOOR SECURITY

The subject of security – embedded security for manufacturing applications to be exact – is complicated, ever changing and very difficult to really understand. Few of us have the background to understand the algorithms, the practices that are used in the IT world, and the complicated concepts and terms that get thrown around.

There is a lot to this technology area, enough for several large books, and this short chapter is just going to highlight the basic concepts and define some of the terms that you need to know to start to have a modicum of understanding of embedded security.

Let's start with the threats. Immediately most of us will think of malicious hackers attacking your control system for fun, or maybe profit. Actually, the biggest threat you have is current and former employees. We've all been surprised in our personal and professional live with the pettiness, vindictiveness and irrationality of people we thought we knew well. When it comes to security, if you can secure your system from internal threats, you are probably going to do a good job with external threats.

What are the risks? Depending on your process and the potential danger to your company and, often, the public, they can be pretty large. They can be as modest as shutting down a minor control system to causing great public harm as in the case of a public utility. You can lose process data, recipes, logic or anything else that might harm your business in various ways. Or you might be exposed to some sort of extortion. The risks in most businesses are too large to ignore.

Is there technology that can rescue you? Well, yes and no. OPC UA has certificate-based security with real teeth. EtherNet/IP is offering an extension called CIP Security and PROFINET IO is developing a security mechanism.

Most of these efforts revolve around the same time of certificate-based security that is used on our corporate networks. Unfortunately, our corporate networks are operating in a homogenous environment where a single standard is dominant, Windows. That's not the case on the factory floor where we have a range of computing power and all sorts of operating systems and software applications. A certificate management problem is thorny, complicated and difficult to support in the Operational Technology (OT) world. Today, there just isn't a good solution.

And the fact that the standards, technologies and products are constantly evolving in this area complicates it even more. Then there are things like legacy systems, organizational practices and company standards that can get in the way. Is the best practice for the St. Louis plant the right approach for Greensboro and San Jose?

You can get a severe migraine just trying to list the issues that you have to confront. Reading the standards can't help as there are many from many different organizations. Here's a few you might consider reading if this subject is of interest to you:

- Cybersecurity Framework Manufacturing Profile by National Institute of Standards and Technology (NIST)

- A series of standards from the ISA/IEC provides a framework to address and mitigate security vulnerabilities. The latest in that series is ISA-62443-4-2, Security for Industrial Automation and Control Systems: Technical Security Requirements for IACS Components, is an adopted standard by both Rockwell Automation and Siemens.

- IEEE 1686-2013 – IEEE Standard for Intelligent Electronic Devices Cyber Security Capabilities; IEEE P1815 – Standard for Electric Power Systems Communications-Distributed

PACKAGING PRODUCT MANAGER'S CONNECTIVITY GUIDE

Network Protocol (DNP3), sponsored by PE/PSCC – Power System Communications and Cybersecurity; and IEEE 1888.3-2013 – IEEE Standard for Ubiquitous Green Community Control Network: Security. All from the Institute of Electrical and Electronic Engineers(IEEE)

As the number of standards might suggest – there is no magic bullet. Vulnerabilities in manufacturing systems are huge threats to financial health and employee safety at all manufacturing companies. That's the bad news but the even worse news is that every customer is going to attack this problem in a different way and technology suppliers must react as these requirements are presented.

INTEGRATING MACHINES WITH PACKML

One of the bigger frustrations' manufacturers face is how to new machines to quickly begin to pay for themselves. The automation world has grown too accepting of the idea that machine integration between vendors of different machines must be complex, laborious and difficult, requiring expensive engineers and incurring significant expense. It's an attitude that can no longer be sustained, in a time when manufacturers face unprecedented strategic challenges in this Cloud-enabled, Big Data, Artificial Intelligence world.

The largest cost for new equipment is now software integration: integration with other machines, with enterprise business systems, and with Cloud applications. Today's integration process is time consuming and exceedingly inefficient. Costs rise when the machine is received with a new or different vendor's programmable controller, a new communication protocol, an unfamiliar HMI, and unique or unusual operating sequences.

PackML is a technology for solving this integration problem. The impetus for the Organization for Machine Automation and Control (OMAC) to develop PackML arose from the dissatisfaction of users and systems integrators frustrated by time, expense and laborious details of integrating control machinery into a coherent system. Integrating a capping machine from one vendor with a labeling machine from another vendor with a sterilizing machine from another is more often a nightmare. Different programming philosophies, control logic, communication protocols, controller platforms and operational states means that each machine requires different operational processes, training, standards and diagnostics methods. Users don't just have a linear increase in complexity with each new packaging component; the complexity increase can be geometric.

PackML is designed to create a consistent look and feel for machinery components integrated into a system. It provides a foundation for vertical and horizontal integration of these machine components irrespective of the vendor, the control system hardware or the specific application. It provides a layer of consistency between vastly different kinds of machines.

By creating a standard set of machine states (Figure 1) and common set of control tags, PackML simplifies the control system development, reduces training and operating costs, and vastly decreases system integration labor and overall expenses.

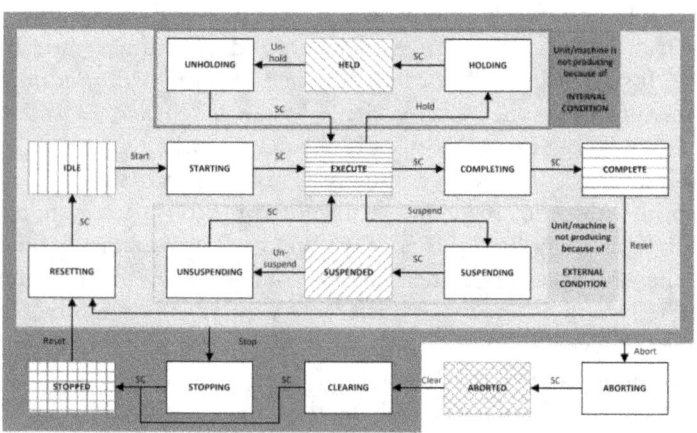

Figure 1 - The PackML State Machine

PackML does not define the specifics of what machine operations occur in any of the machine states it defines. For example, it specifies the transitions that move a machine into Starting State or the Idle State, but it does not specify the functionality of those states.

PACKAGING PRODUCT MANAGER'S CONNECTIVITY GUIDE

However, by having a set of common states and control tags, status tags and administration tags, the monitoring of any particular machine is identical to monitoring every other PackML machine. This lowers maintenance, support and training costs.

PackML can be compared to the generic object notation of EtherNet/IP, ProfiNet IO or BACnet, although its application-level functionality vastly exceeds what's available in other systems. PackML is a standard for modeling machine behavior that provides a standard mechanism for monitoring and understanding industrial machine operation. PackML decreases the integration effort required to exchange data between machines and between users and machines via operator interfaces.

PackML models machine data through its state machine and the use of standard PackML Tags but does not specify how those tags get from one machine to another or from a machine to an HMI. It does not specify any transports, security, encodings, interfaces or physical media. Technologies like OPC UA and Ethernet TCP/IP provide that.

A system implementing this is illustrated in Figure 4. In this packaging system all machines support PackML over OPC UA. Because all machines use standard PackML tags over OPC UA, it is much less complex to configure other downstream machines to properly handle faults, alarms, starts, stops and other process status. And HMIs can be much more easily constructed from the standard state machines and tags used by all machines.

PACKAGING PRODUCT MANAGER'S CONNECTIVITY GUIDE

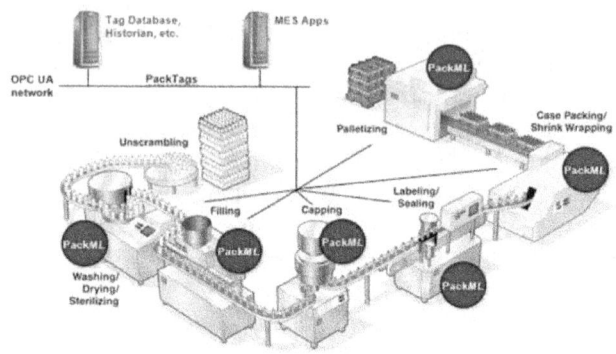

Figure 2 - PackML Equipped Manufacturing Cell

Contrast that with a situation where every machine in this system is built with a proprietary state machine, with a set of proprietary tags encoded in some binary format with its unique security mechanism. Imagine if each machine uses different tag names for state information, different tags for diagnostics, and different tag names for data values. Imagine if some of the machines use XML while others use their own binary data encoding. Imagine if some only use HTTPS for security while others use a user-password security scheme. Imagine being responsible for integrating that system. Building a system from those components would be a time and expense nightmare.

Unfortunately, in the automation world, we've grown too accepting of the idea that machine integration between vendors of different machines must be complex, laborious and difficult, requiring expensive engineers and incurring significant expense. With adherence to the kinds of standards described in this paper, system integration for highly connected solutions will cease being complex and expensive. Instead it will just work.

The beauty of this approach is that it goes beyond packaging lines. These concepts are just as relevant to machinery operations in diaper and tissue converting, food and beverage, automobile production and many other industries. The PackML concepts, tags and state machines can be used in any of these industries and more.

PACKAGING PRODUCT MANAGER'S CONNECTIVITY GUIDE

It wouldn't be fair to say that OPC UA and PackML are the only two magic bullets that can solve all manufacturers strategic challenges in this Cloud-enabled, Big Data, Artificial Intelligence world. But OPC UA and PackML are proven technologies that reduce integration costs, increase productivity and vastly simplify how you get data into your enterprise or Cloud applications. Getting the data we need to operate more productively and efficiently is the new job – and leveraging OPC UA and application layer architectures like PackML are going to be some of the tools you'll need in the future!

PACKAGING PRODUCT MANAGER'S CONNECTIVITY GUIDE

IF YOU STILL USE ASCII...

Today there are still industrial devices around that use older style ASCII communications. In the days before Modbus, custom ASCII communications was very popular. Sometimes the commands would be as simple as "$GO" to start a drive but sometimes there were complicated with multiple parameters such as
"W3 52,175.6,.3345"

Custom ASCII was found in everything from drives to chart recorders to weigh scales and barcode readers. And, over time, some of those ASCII command sets expanded into extensive command lists. With the unsophisticated microprocessors of the 80s, 90s and 2000s and before standard communication technologies, that was a very reasonable way to do things.

Now, of course, that seems awfully ancient. There's no argument at how outdated these devices are but in manufacturing, the requirements haven't changed. An ASCII weigh scale, for example, still calculates weights as well as it always did. That means that these old devices are still functional, they still work well and customers still want to use them. Just like the old TVs, they're not going away. Manufacturers and system integrators need to deal with them. The problem that they have is that they just don't integrate well into today's network architectures.

If you have one of these ASCII devices in your product line, there are things you can do to make your device more easily integrated into

PACKAGING PRODUCT MANAGER'S CONNECTIVITY GUIDE

today's manufacturing system. There are two ideas you might consider that don't require you to reengineer your entire device; 1) Adapt your current ASCII protocol into something more palatable to today's control system or 2) Convert it to a more functional protocol technology.

Adapting your protocol means using a device that can simply convert your ASCII commands into something like JSON or XML (see earlier chapter). JSON is an object oriented, ASCII version of XML. Where XML uses tags of the format <Temp>52</Temp>, JSON uses curly braces: {"name":"Emily", "age":34, "car":null}. The advantage to JSON is IT devices using Java can easily convert JSON into memory variables that can be logically processed by code.

Converting your protocol means using a device that makes your ASCII device look like a Modbus TCP Slave, an EtherNet/IP Adapter or PROFINET IO end device. To do that, some subset of your ASCII command set must be mapped into the base data representation of that protocol.

Either approach can extend the life of your current ASCII device whether you have the intention of replacing it in the future or you just want to continue selling it as long as you can.

Neither one of these approaches is difficult but they do require some sort of hardware and software to make that happen.

PACKAGING PRODUCT MANAGER'S CONNECTIVITY GUIDE

ADDENDUMS

PACKAGING PRODUCT MANAGER'S CONNECTIVITY GUIDE

GLOSSARY

Acyclic Communication Acyclic communication messages are used to transfer commands and information between programmable controllers and end devices on an as-needed basis.

Application Layer The application layer is also known as the user software layer. This is the user application software that converts input signals into digital inputs and converts digital outputs into output signals. The application layer communicates to the industrial protocol over some host interface.

ARP Address Resolution Protocol (ARP) is a component of TCP that identifies the hardware MAC (Media Access Control) address of a device from a TCP/IP address.

ASCII ASCII is a data format that is universally decodable by any computer on the planet. It is a mechanism for representing letters, numbers and

Bandwidth special characters in computer memory.
Bandwidth is the capacity of the network to move data. A 50 megabit network can move 50 megabits of data per second, but that doesn't mean that your device is getting all that capacity. If there is another device on the network, bothcan get 25 megabits of capacity. If there are ten devices, assuming each gets equal time, each can get 5 megabits of capacity per second and so on.[1] Even though any individual message is moving very fast (maybe 1 Gbps), your device only gets access to that very fast pipe intermittently. How little access you get is your bandwidth.

CAN BUS CAN BUS is a communication standard adopted in the 1990s that provided for fast, low cost communication between IO devices that have small numbers of I/O points. CAN is a standard technology used in brands of automobiles.

Client A Client is a device that makes connections to end devices on an industrial network. A Client device is unlike a client in an IT (Internet Technology) network where a client is more of a remote and or edge network device. In an industrial network, a client is usually a programmable

---

[1] Note that I am simplifying here. There are other issues that result in some devices getting more of the available bandwidth than other devices.

PACKAGING PRODUCT MANAGER'S CONNECTIVITY GUIDE

Cyclic Communication	controller that makes connections to end devices. Cyclic communication messages used to transfer sets of inputs and outputs between programmable controllers and end devices. It is a communication method in which messages are transmitted on a continual basis. There is no relationship or dependency between an outgoing cyclical message and an incoming cyclical message.
Ethernet	Ethernet is a computer networking technology standard for local area networks (LANs). Ethernet device communicate by dividing a stream of data into individual packets called frames. Each frame contains source and destination addresses and error-checking data so that damaged data can be detected and retransmitted.
Latency	You can think of latency as another word for delay. When we talk about latency with an industrial network like EtherNet/IP, PROFINET IO or even Modbus TCP, we're talking about the delay accumulated as the packet makes its way from the Master device (EtherNet/IP Scanner, PROFINET IO Controller or Modbus TCP Master) to the Slave device and back to the Master. There can be delays in gateways, routers and in the end device itself. If the Slave device takes 50 msecs to respond to an acyclic request, that's an additional 50 msecs of delay added to all the other delays for a request of a controller to be satisfied

PACKAGING PRODUCT MANAGER'S CONNECTIVITY GUIDE

Network Ping	by a response from a target device. A network ping is a protocol component of TCP (Transport Control Protocol) that measures and sends messages to modes and reports the time delay in receiving a response. It can be used to determine if a node on an Ethernet network is reachable.
Object Model	An object model is a digital representation of a device's data in a specific format. Every technology supports an object model but the implementation mechanism is very different. In OPC UA, the object model implies a set of hierarchical relationships between objects, nodes and variables. In EtherNet/IP, the object model is a one-level set of objects and attributes. In PROFINET IO, the object model is a rack, slot, module, and IO point.
Programmable Controller	An industrial computer that is ruggedized for the factory environment. A programmable controller incorporates three required functions; 1) collecting digital inputs from external devices which convert real world signals to digital data, 2) processing control logic which operates on those inputs and 3) sending digital outputs to external devices which convert that digital data to real world analog signals.
Server	A Server is an end-device on an industrial network that converts real world analog signals (temperature, rotational speed...etc.) to digital inputs

PACKAGING PRODUCT MANAGER'S CONNECTIVITY GUIDE

and digital outputs to real world analog signals that modify some machine state.

Speed Network speed strictly refers to the number of bits we can get through a wire in one second. You can also think of it as the time a bit is present on a wire. At very slow baud rates, like 300 bits per second, the bit is present on the wire for about 3 milliseconds. At 1 meg (1,000,000 bits per second), the bit is on the wire for about 1 microsecond. Network speeds are measured in bits per second (bps), kilobits or 1,000 bits per second, megabits or 1,000 kilobits per second (Mbps) and gigabits or 1,000 megabits per second (Gbps).

TCP/IP A set of communication protocols that supports the operation of Ethernet. This set of protocol components is known as a stack and often is part of an operating system like Linux or Windows 10 but can also be purchased from a software supplier.

Throughput Throughput is the actual amount of data that is transferred across a network link. A network link may be rated for 50 Mbps, but the actual amount of data it can transfer may be limited to something much less. Throughput is reduced by the number of devices on the network or network segment, the protocol being used and many other factors. A common way to visualize this is as a congested three-lane highway. It has the capacity to move a car at 90 mph with no traffic. But at peak traffic times or when the weather is bad or when an officer is writing a ticket, the cars move at slower speeds. That's how throughput gets degraded. Any number of things can cause the throughput on your network to degrade

ABOUT THE AUTHOR

John S Rinaldi, is Chief Strategist, Business Development Manager and CEO of Real Time Automation (RTA) in Pewaukee, WI.

After escaping from Marquette University with a degree in Electrical Engineering (graduating cum laude, no less), John worked in various jobs in the Automation Industry before fleeing back to the comfortable halls of academia. At the University of Connecticut, he once again talked his way into a degree, this time in Computer Science (MS CS).

John achieved marginal success as a Control Engineer, a Software Developer and IT Manager before founding Real Time Automation because "long term employment prospects are somewhat bleak for loose cannons."

PACKAGING PRODUCT MANAGER'S CONNECTIVITY GUIDE

With a strong desire to avoid work, responsibility and decision making, John had to build a great team at Real Time Automation. And he did. RTA now supplies network converters for industrial and building automation applications all over the world. With a focus on simplicity, US support, fast service, expert consulting, and tailoring for specific customer applications, RTA has become a leading supplier of gateways worldwide.

John freely admits that the success of RTA is solely attributable to the incredible staff that like working for an odd, quirky company with a single focus: "Create solutions so simple to use that the hardest part of their integration is opening the box."

John is a recognized expert in industrial networks and a speaker. He's spoken at events sponsored by the ODVA, Profinet International, the OPC UA Foundation and many industrial distributor events in the US and Europe. John is an author of seven books, hundreds of blog articles and scores of articles on communications and control in Automation magazine, Control Magazine and many others. John is a published article with five books on Industrial Networking.

John is famous for the articles in his Automation newsletter, one of the most widely read newsletters in the automation industry. The newsletter is a combination of John's opinions, technical insights, humor and fun. It's a must read for many automation professionals.

You can reach John here:

John S Rinaldi
Real Time Automation
N26 W23315 Paul Rd
Pewaukee, WI 53072
jrinaldi@rtautomation.com
262-436-9299 (Office)
414-460-6556 (Cell)

http://www.rtautomation.com/contact-us/
https://www.linkedin.com/in/johnsrinaldi

PACKAGING PRODUCT MANAGER'S CONNECTIVITY GUIDE

> "Real Time Automation's expertise with Industrial Ethernet protocols, PLCs, and Industrial Computer solutions proved invaluable to us and provided us the ability to rapidly migrate our products to state of the art industrial control platforms. . They have been accommodating in incorporating features that had not been anticipated early on in the specification process. The process of working with RTA's engineers has been as efficient as working with our own engineers in remote locations. They have assisted us in the definition, development, documentation and certification of a new line of products for the Industrial Ethernet markets within a relatively short schedule. I would recommend their services and products to other companies requiring their expertise."
>
> Brian Monahan
Research & Development Manager
Escort Memory Systems

PACKAGING PRODUCT MANAGER'S CONNECTIVITY GUIDE

OTHER BOOKS BY JOHN RINALDI

PACKAGING PRODUCT MANAGER'S CONNECTIVITY GUIDE

"EtherNet/IP: The Everyman's Guide"

The Everyman's Guide to EtherNet/IP is a easy-to-understand guide to one of the most widely used protocol in manufacturing. The book explains CIP, the Common Industrial Protocol, and the core of EtherNet/IP, DeviceNet, CompoNet and ControlNet. With this handy guide on your bedside table, you'll be able to navigate the world of EtherNet/IP and, who knows, maybe entertain a companion with interesting tidbits. Dip into any chapter, in any order, and you'll find plenty of clear explanations, with tables and diagrams, to guide you through EtherNet/IP.

http://rtabooks.com/ethernetip

PACKAGING PRODUCT MANAGER'S CONNECTIVITY GUIDE

"OPC UA: The Basics"

This book is a quick introduction to OPC UA for people who don't need to become experts but would like to talk knowledgably about OPC UA technology. It's not very detailed and in a few places mischaracterizes the technology, but it accomplishes its purpose as a basic introduction to the technology.

http://rtabooks.com/opcuabasics

PACKAGING PRODUCT MANAGER'S CONNECTIVITY GUIDE

"OPC UA: The Everyman's Guide"

This is book for the dedicated professional that needs to really understand OPC UA technology. It's organized well and gives the reader what they need to know about OPC UA without losing them in the details. In fact, the book organizes the technical topics in such a way that the reader can choose to read just a quick overview of each topic, the concepts you need to know about that topic, or the full details.

http://rtabooks.com/opcua

PACKAGING PRODUCT MANAGER'S CONNECTIVITY GUIDE

"Modbus – The Everyman's Guide to Modbus"

In manufacturing automation, we use a lot of old technology. Yet even in our world, Modbus isn't just old technology. IT'S ANCIENT TECHNOLOGY. We live in a new age. The age of enterprise communications. It's an age where automation and the factory floor are changing in ways that weren't imaginable just a few short years ago. And despite all this, Modbus is still with us and is going to be with us for a long time. Modbus devices have permeated every kind of automation and will continue to over the next hundred years due to their simplicity and because Modbus is perfect for a lot of simple devices.

This book describes Modbus technology and the role that Modbus will continue to play in the future.

http://rtabooks.com/modbus

PACKAGING PRODUCT MANAGER'S CONNECTIVITY GUIDE

"INDUSTRIAL ETHERNET"

This book is an introduction to Industrial Ethernet. This book is considered the go-to guidebook for people who need to fully understand factory floor Ethernet and for those who need to have a basic understanding of Ethernet and TCP/IP terminology, Ethernet hardware, Ethernet software, Ethernet security, and the Internet of Things (IoT).

It includes practical reference charts and installation, maintenance, troubleshooting, and security tips, which make this book an ideal quick reference resource at project meetings and on the job.

http://rtabooks.com/industrialethernet

www.ingramcontent.com/pod-product-compliance
Lightning Source LLC
Chambersburg PA
CBHW070459220526
45466CB00004B/1893